JOB HAZARD ANALYSIS (JHA) FORMS

A great tool for ensuring you and your co-workers go home safely at the end of each shift. Nothing should compromise safety.

Image Credits: Royalty free images reproduced under license from stock image repositories.

HOW TO USE THIS BOOK:

REFER TO THE SAMPLE PAGE TO HELP YOU FILL OUT THE JHA FORM.

SAMPLE JHA

Job Title: Change Tyre on vehicle
Date: 4/12/2018
Supervisor: John Brown

JOB STEPS, HAZARDS AND CONTROLS

Job Step	Hazards	Controls
Make area safe	Oncoming traffic	Safety cones
Apply jack	Vehicle moves	Apply chocks
Prepare spare tyre	Back sprain	Manual handling techniques
Replace tyre	Loose nuts	Check nuts

SIGNATURES

Name	Signature
Harry Smith	*Harry Smith*
Steve Jones	*Steve Jones*

JHA

Job Title:

Date:

Supervisor:

JOB STEPS, HAZARDS AND CONTROLS

Job Step   Hazards   Controls

SIGNATURES

Name   Signature

JHA

Job Title:

Date:

Supervisor:

JOB STEPS, HAZARDS AND CONTROLS

Job Step	Hazards	Controls

SIGNATURES

Name   Signature

JHA

Job Title:

Date:

Supervisor:

JOB STEPS, HAZARDS AND CONTROLS

Job Step   Hazards   Controls

SIGNATURES

Name   Signature

JHA

Job Title:

Date:

Supervisor:

JOB STEPS, HAZARDS AND CONTROLS

Job Step   Hazards   Controls

SIGNATURES

Name   Signature

JHA

Job Title:

Date:

Supervisor:

JOB STEPS, HAZARDS AND CONTROLS

Job Step	Hazards	Controls

SIGNATURES

Name   Signature

JHA

Job Title:

Date:

Supervisor:

JOB STEPS, HAZARDS AND CONTROLS

Job Step   Hazards   Controls

SIGNATURES

Name   Signature

JHA

Job Title:

Date:

Supervisor:

JOB STEPS, HAZARDS AND CONTROLS

Job Step   Hazards   Controls

SIGNATURES

Name   Signature

JHA

Job Title:

Date:

Supervisor:

JOB STEPS, HAZARDS AND CONTROLS

Job Step	Hazards	Controls

SIGNATURES

Name   Signature

JHA

Job Title:

Date:

Supervisor:

JOB STEPS, HAZARDS AND CONTROLS

Job Step   Hazards   Controls

SIGNATURES

Name   Signature

JHA

Job Title:

Date:

Supervisor:

JOB STEPS, HAZARDS AND CONTROLS

Job Step   Hazards   Controls

SIGNATURES

Name   Signature

JHA

Job Title:

Date:

Supervisor:

JOB STEPS, HAZARDS AND CONTROLS

Job Step   Hazards   Controls

SIGNATURES

Name	Signature
.	

JHA

Job Title:

Date:

Supervisor:

JOB STEPS, HAZARDS AND CONTROLS

Job Step   Hazards   Controls

SIGNATURES

Name   Signature

JHA

Job Title:

Date:

Supervisor:

JOB STEPS, HAZARDS AND CONTROLS

Job Step   Hazards   Controls

SIGNATURES

Name   Signature

JHA

Job Title:

Date:

Supervisor:

JOB STEPS, HAZARDS AND CONTROLS

Job Step   Hazards   Controls

SIGNATURES

Name   Signature

JHA

Job Title:

Date:

Supervisor:

JOB STEPS, HAZARDS AND CONTROLS

Job Step   Hazards   Controls

SIGNATURES

Name   Signature

JHA

Job Title:

Date:

Supervisor:

JOB STEPS, HAZARDS AND CONTROLS

Job Step   Hazards   Controls

SIGNATURES

Name   Signature

JHA

Job Title:

Date:

Supervisor:

JOB STEPS, HAZARDS AND CONTROLS

Job Step   Hazards   Controls

SIGNATURES

Name   Signature

JHA

Job Title:

Date:

Supervisor:

JOB STEPS, HAZARDS AND CONTROLS

Job Step   Hazards   Controls

SIGNATURES

Name   Signature

JHA

Job Title:

Date:

Supervisor:

JOB STEPS, HAZARDS AND CONTROLS

Job Step   Hazards   Controls

SIGNATURES

Name   Signature

JHA

Job Title:

Date:

Supervisor:

JOB STEPS, HAZARDS AND CONTROLS

Job Step	Hazards	Controls

SIGNATURES

Name   Signature

JHA

Job Title:

Date:

Supervisor:

JOB STEPS, HAZARDS AND CONTROLS

Job Step	Hazards	Controls

SIGNATURES

Name   Signature

JHA

Job Title:

Date:

Supervisor:

JOB STEPS, HAZARDS AND CONTROLS

Job Step   Hazards   Controls

SIGNATURES

Name   Signature

JHA

Job Title:

Date:

Supervisor:

JOB STEPS, HAZARDS AND CONTROLS

Job Step	Hazards	Controls

SIGNATURES

Name   Signature

JHA

Job Title:

Date:

Supervisor:

JOB STEPS, HAZARDS AND CONTROLS

Job Step   Hazards   Controls

SIGNATURES

Name   Signature

JHA

Job Title:

Date:

Supervisor:

JOB STEPS, HAZARDS AND CONTROLS

Job Step   Hazards   Controls

SIGNATURES

Name   Signature

JHA

Job Title:

Date:

Supervisor:

JOB STEPS, HAZARDS AND CONTROLS

Job Step	Hazards	Controls

SIGNATURES

Name   Signature

JHA

Job Title:
Date:
Supervisor:

JOB STEPS, HAZARDS AND CONTROLS

Job Step	Hazards	Controls

SIGNATURES

Name   Signature

JHA

Job Title:

Date:

Supervisor:

JOB STEPS, HAZARDS AND CONTROLS

Job Step	Hazards	Controls

SIGNATURES

Name   Signature

JHA

Job Title:

Date:

Supervisor:

JOB STEPS, HAZARDS AND CONTROLS

Job Step   Hazards   Controls

SIGNATURES

Name   Signature

JHA

Job Title:

Date:

Supervisor:

JOB STEPS, HAZARDS AND CONTROLS

Job Step   Hazards   Controls

SIGNATURES

Name   Signature

JHA

Job Title:

Date:

Supervisor:

JOB STEPS, HAZARDS AND CONTROLS

Job Step   Hazards   Controls

SIGNATURES

Name	Signature

JHA

Job Title:
Date:
Supervisor:

JOB STEPS, HAZARDS AND CONTROLS

Job Step   Hazards   Controls

SIGNATURES

Name   Signature

JHA

Job Title:	
Date:	
Supervisor:	

JOB STEPS, HAZARDS AND CONTROLS

Job Step	Hazards	Controls

SIGNATURES

Name   Signature

JHA

Job Title:

Date:

Supervisor:

JOB STEPS, HAZARDS AND CONTROLS

Job Step   Hazards   Controls

SIGNATURES

Name   Signature

JHA

Job Title:

Date:

Supervisor:

JOB STEPS, HAZARDS AND CONTROLS

Job Step   Hazards   Controls

SIGNATURES

Name   Signature

JHA

Job Title:

Date:

Supervisor:

JOB STEPS, HAZARDS AND CONTROLS

Job Step   Hazards   Controls

SIGNATURES

Name   Signature

JHA

Job Title:

Date:

Supervisor:

JOB STEPS, HAZARDS AND CONTROLS

Job Step	Hazards	Controls

SIGNATURES

Name   Signature

JHA

Job Title:

Date:

Supervisor:

JOB STEPS, HAZARDS AND CONTROLS

Job Step	Hazards	Controls

SIGNATURES

Name   Signature

JHA

Job Title:

Date:

Supervisor:

JOB STEPS, HAZARDS AND CONTROLS

Job Step   Hazards   Controls

SIGNATURES

Name   Signature

JHA

Job Title:

Date:

Supervisor:

JOB STEPS, HAZARDS AND CONTROLS

Job Step	Hazards	Controls

SIGNATURES

Name   Signature

JHA

Job Title:

Date:

Supervisor:

JOB STEPS, HAZARDS AND CONTROLS

Job Step   Hazards   Controls

SIGNATURES

Name   Signature

JHA

Job Title:

Date:

Supervisor:

JOB STEPS, HAZARDS AND CONTROLS

Job Step | Hazards | Controls

SIGNATURES

Name   Signature

JHA

Job Title:

Date:

Supervisor:

JOB STEPS, HAZARDS AND CONTROLS

Job Step   Hazards   Controls

SIGNATURES

Name   Signature

JHA

Job Title:

Date:

Supervisor:

JOB STEPS, HAZARDS AND CONTROLS

Job Step   Hazards   Controls

SIGNATURES

Name   Signature

JHA

Job Title:

Date:

Supervisor:

JOB STEPS, HAZARDS AND CONTROLS

Job Step   Hazards   Controls

SIGNATURES

Name   Signature

JHA

Job Title:

Date:

Supervisor:

JOB STEPS, HAZARDS AND CONTROLS

Job Step	Hazards	Controls

SIGNATURES

Name   Signature

JHA

Job Title:

Date:

Supervisor:

JOB STEPS, HAZARDS AND CONTROLS

Job Step   Hazards   Controls

SIGNATURES

Name   Signature

JHA

Job Title:

Date:

Supervisor:

JOB STEPS, HAZARDS AND CONTROLS

Job Step   Hazards   Controls

SIGNATURES

Name   Signature

JHA

Job Title:

Date:

Supervisor:

JOB STEPS, HAZARDS AND CONTROLS

Job Step   Hazards   Controls

SIGNATURES

Name   Signature